Mohamed Saleh

Pruning and Population Adjustment in Relation to Fruit Trees Productivity

Pruning and spacing of Fruit trees

GRIN Publishing

Bibliographic information published by the German National Library:

The German National Library lists this publication in the National Bibliography; detailed bibliographic data are available on the Internet at http://dnb.dnb.de .

Imprint:

Copyright © 2002 GRIN Verlag, Open Publishing GmbH
Print and binding: Books on Demand GmbH, Norderstedt Germany
ISBN: 978-3-640-97601-0

This book at GRIN:

http://www.grin.com/en/e-book/176054/pruning-and-population-adjustment-in-relation-to-fruit-trees-productivity

PRUNING AND POPULATION ADJUSTMENT IN RELATION TO FRUIT TREES PRODUCTIVITY

A REVIEW ARTICLE

BY

MOHAMED MAHER SAAD SALEH
Prof. of Pomology

Pomology Research Department
National Research Center

CONTENTS

ABSTRACT

High planting density is considered one of the modern management systems in new fruit orchards. Pruning and population adjustment are some of the main way to approach high productivity and good fruit quality under the modern management systems.

Pruning is defined as the management of plant structure and fruiting wood and it is considered one of the main factors affecting fruit trees production. There are many objectives for pruning i.e. controlling the direction of growth, developing a strong framework, controlling the amount of growth, improved productiveness, improving quality product, utilizing space efficiently and increasing the usefulness of plant. There are three main systems of fruit trees training (i) Central leader system, (ii) Modified leader system, (iii) Open center or vase system. Also there are several kinds of pruning i.e. heading back and thinning out, fine and coarse pruning and root pruning. However, it is important to select the suitable training system and kind of pruning according to the target of the pruning to reach optimum vegetative growth and maximum yield with high fruit quality.

On the other hand, population adjustment (plant distribution or planting density) is very important factor affecting fruit trees production. We can reach the optimum usefulness of the planted area by using best population adjustment of trees in the plantation. Trees can be planted using different systems:- triangle, rectangle, square system…..etc. On the other side, presence of pollinator trees through the orchard is very important for many fruit trees. Pollinator can be planted either in (1) Complete along rows, as one row of pollinator per two or four rows of the main cultivar. (2) The second tree in the second row. (3) The third tree in the third row. (4) Across rows through the orchard. On the other hand, good plant distribution is very helpful for mechanical serves in the plantation,

We can say that the key objective in planning a new orchard should be to maximize yield in the early years and still effectively produce larger yield of high quality fruits in the next years. This may be happened by utilizing space efficiently and increasing the usefulness of plant using the suitable training system, kind of pruning, also good population adjustment of the trees in the orchard beside the importance of the other factors affecting production process.

<u>INTRODUCTION</u>

There are many factors affecting fruit trees production, some of these factors are related to the environment conditions and the others are related to the plant it self.

Pruning of fruit trees and population adjustment play an important limiting role on fruit trees productivity.

Pruning is considered the main factor affecting some fruit trees production such as grapevine; also it is an essential factor for the productivity of either deciduous or evergreen fruit trees.

Population adjustment of fruit trees plays very important role as a one of the first steps of plant arrangement in the plantation, since plant distribution or planting density (the same meaning of population adjustment) are generally affecting fruit trees production, and had a great effect in a special case such as banana or trees which need a pollinator.

However, it is clear that there is a correlation between both pruning and population adjustment and both are considered limiting factors affecting fruit trees productivity. In this respect, **Zimmerman and Steffens (1995)** reported that tree size was inversely related to planting density.

So, we will try to make a focus on the role and the effect of both pruning and population adjustment on fruit trees through the following items.

1.　　Pruning

The management of plant structure and fruiting wood is called pruning. It involves removing parts of a plant's top or root system to increase its usefulness. Limbs, branches, twig, shoot, or roots can be removed. Pruning also includes the training of plants, or shaping them to forms that function more efficiently. Pruning is important for the successful production of both deciduous or evergreen fruit trees. Horticulturists look at yield, size, colour, shape, or quality of fruits in terms of potential profit, while for amateurs factors like size, beauty, or quality are ends in themselves. Pruning helps both to achieve their goals more effectively.

1.1　　Principles of pruning

Some important principles of pruning are summarized, following **Malik,(2000)** :-

1.1.1. Modification of apical dominance:-

Apical dominance occurs when hormones produced in the stem apices travel down the stem and inhibit or reduce branching and growth of lateral buds. When the terminal growing point is removed, the production and flow of these hormones to lateral buds is stopped and the initiation rate of lateral growth of branches is increased.

1.1.2   Balance of roots and top:-

Plant growth, development, and reproduction are influenced by the ratio of roots to top. Reduction of leaf area in the growing season or reduction of the number of buds in the dormant season has little effect on root area, but a reduced number of growing points results in stronger shoots with larger leaves. Increase leaf area increases transpiration and photosynthesis, and puts more demands on the roots. However, root pruning reduces the absorbing area, slowing top growth. Stored food is utilized to replace roots, and top growth dose not resume immediately after root replacement since manufactured food must first be stored in the stem. Proper pruning affects a balance of top and roots.

1.1.3. Altering growth phases:-

Regular annual pruning of a growing tree stimulates shoot growth. Heavy annual pruning of young fruit trees delays early fruit production, therefore pruning should be minimal from the juvenile to the productive stage. For maximum flower and fruit production, however, a plant should show good annual shoot growth. If annual shoot growth decreases, as happens with older trees, pruning will stimulate growth and production is usually increased. However, excessive pruning can cause the plant to a vegetative state.

1.1.4. Environmental factors:-

Desirable pruning and training practices are influenced by several environmental factors:-

- Trees grown in heavy shade are pale coloured, have fewer flowers, and are usually smaller. Frequent pruning to maintain good form allows light to strike the leaves and produces dense foliage in hedges.
- Pruning can also influence air movement. In spreading foliage, air movement is increased. Open structures allow better spray penetration for controlling insect pests and diseases.

- Excess moisture tends to produce water sprouts on trunks and primary branches. If excess moisture is preceded or accompanied by severe pruning, the increased water sprout production wastes plant growth.

- Since pruning reduce transpiration, it is useful during drought periods.

- Temperature should be considered when deciding whether to prune. Soft, succulent growth resulting from over- pruning or late summer pruning is more susceptible to winter injury, because there is less time for hardening and storing food before the cold weather.

- Crown and crotch injuries due to cold are more likely on trees, which have not been trained or pruned to desirable branching angles.

- The trunks of trees trained to a low-headed shape receive less intense light, and the bark may be protected from sunscald.

1.2   Objectives of pruning

Major objectives of pruning are summarized below as **Malik, (2000)** :-

1.2.1. Controlling the direction of growth:-

The natural form of a plant can be modified to induce it to branch and spread more profusely. Low-branching types can be trained to branch higher. Branches can also be trained to grow away from utility wires or buildings.

1.2.2. Developing a strong framework:-

Some trees have naturally narrow crotch angles (40° or less from the vertical). Narrow crotches result in a greater loss of limbs from windstorms and heavy loads of fruit than crotches with larger angles. The strongest crotches are those in which branches grow up from the trunk at angles ranging from $40 - 90^{\circ}$. Scaffold branches should be evenly spaced around the tree, with each branch at least 90° from the next one.

1.2.3 Controlling the amount of growth:-

Pruning can either dwarf or invigorate a tree. The type, manner, and time of pruning is determined by the objective. A combination of dormant and summer pruning promotes dwarfness. It is the frequency of pruning rather than its severity that is critical in promoting dwarfness. Increased in vigour is produced by dormant pruning of older wood.

1.2.4. Improving productiveness:-

Yield can be either increased or decreased by pruning. The type of pruning depends on the fruit-bearing habits or the vegetative response of each species. Decreasing the number of fruit buds will usually give fewer but larger fruits and may increase the percentage of desirable fruit. Severe pruning can stimulate excess vegetative growth of scion wood, but suppress fruit bearing. In this respect, Flam seedless and Ruby seedless grapes were trained using three systems, head, cane and cordon. However, results indicated that the percentage of bursted buds were affected by training system, since it was higher in cordon system, lower in cane pruning, the percentage of fruitful buds did not affected. In Flam seedless, bunch number/vine was higher for both cordon and dead training system, while for Ruby seedless, vines trained according to the cordon system beard the highest yield per vine and the lower was the cane system. There were no differences for various fruit quality parameters as an effect of training system for both cultivars, **Gaser and Osman (1999).**

1.2.5. Improving quality of product:-

Fruit has better colour and flavour with adequate light. Improving product quality is one of the main reasons for annual dormant pruning of many fruit trees.

1.2.6. Utilizing space efficiently:-

Training, staking, and pruning are done simultaneously when trees are grown for the most efficient utilization of space. Pruning facilitates cultural operations, for example good spray penetration and the use of maintenance equipment, and also makes fruit picking easier. Fruit trees with very tops are difficult to harvest.

1.2.7. Increasing the usefulness of plants:-

Pruning to modify growth increases plant utility. A properly pruned shade tree can provide heavier shade and a regularly clipped hedge can become almost impenetrable. Such trees and shrubs provide better screening and wind protection. Vines can produce more concealing foliage, and large or spreading plants can be miniaturized by training and pruning.

1.2.8 Controlling diseases:-

Pruning may be useful in controlling some plant diseases such as fire blight in apple and pear trees. In this respect, **Shaffer (2000)** reported that since the fire blight bacteria winter in infected plant parts, the removal of these parts by pruning can reduce the severity of fire blight the following year. On the other hand, **Retamales *et.al.*(2000)** studied the effect of crop loads of 55, 70 or 150 fruits/tree on Bitter pit disease in Braeburn/seedling apple trees. Results showed that Bitter pit incidence was 50% with 150 fruits/tree compared with 75% with 55 fruits/tree.

1.3   Training

Training is pruning management done to develop a tree framework strong enough to bear large fruit crops without the branches breaking. There are three main training system as **Malik, (2000)** description: -

- Central leader,   - Modified leader,   - Open center or vase   system.

1.3.1. Central leader system:-

This system resembles the natural growth pattern of most trees. Trees are trained to a main stem and a series of well-spaced, subordinate lateral branches. Apical growth is encouraged, resulting in taller trees than with other systems. The advantage of this system is the development of strong crotches, and the disadvantage is internal shading, which may weaken the central leader and thus reduce the life of the tree.

1.3.2. Modified leader system:-

This system reduces the height of the main trunk and encourages scaffold branches to become larger and longer, thus lowering the top of the tree. It combines features of the central and open center systems. The central leader is cut back slightly so that it does not become dominant, and laterals are cut back and selected repeatedly until an appropriate number and distribution of branches is reached. The central leader is then cut and the tree is left with a rounded open top, well-spaced limbs, a strong framework, and well-distributed fruiting wood. It is low enough to facilitate various orchard operations. This is the most desirable pruning system for many fruits.

1.3.3. Open center or vase system:-

This training system develops a series of well-spaced, coordinate lateral branches rather than a main or central trunk. These branches are cut back equally each year, which gives them equal dominance. The advantages of this system are sufficient light penetration for the fruiting of inner branches, and a low-headed tree that facilitates pruning, thinning, spraying, and picking. Its main disadvantage is that the tree becomes weak with crowded crotches, which often break under a heavy load of fruit.

In this respect, Apple trees when planted at high-density orchard systems (greater than 500 trees per acre), such as the vertical axis system, can be used to produce fruit in the second or third year after planting. In contrast, low-density orchard system (150 to 200 trees per acre) generally requires five years to come into production, **Michele (1997).** On the other hand, early ripening peach "Flordaprince" gave higher fruit yield, greater leaf area and accumulated more dry mass in above- ground components per tree when trained using central leader system comparing with Y shape trees system, **Caruso *et.al.*(1999).**

1.4 Kinds of pruning

There are several different kinds of pruning, involving both top and root.

- Heading back and thinning out.
- Fine and coarse (bulk) pruning.
- Root pruning.

There are also a number of special pruning practices. Fruit thinning is also considered to be a type of pruning. Desirable results are often accomplished through a combination of these methods, which are explained in detail below:-

1.4.1 Thinning out and heading back:-

Thinning out means to remove certain shoots, canes, spurs, or branches completely from the base, while heading back means removing only the terminal portions of these parts. Thinning out may remove the same percentage of plant's structure as heading back, but there is substantial difference in what is removed, and an even greater difference in the response of the plant. Heading back leaves the shoots more crowded than does a corresponding thinning. Thinning cut usually does not effect much reduction in height or spread, but heading back of comparable severity effects substantial reduction in height, spread, or both, and leaves a less open plant. Thinning plants become taller, and more spreading and open, while headed plants become more compact. In general, heading back stimulates the development of more growing points by overcoming apical dominance, while thinning out does not have this effect.

1.4.2 Fine and coarse (bulk) pruning:-

As bearing branches become older and weaker, they set fewer flowers and fruit of poor quality, and die within a few months. Removing these branches earlier will result in vigorous new wood. Removal involves a large number of small cuts, it is fine or thin-wood pruning. The term applies to shoots, canes, spurs, and older growth. Coarse or bulk pruning is done from the upper portions of the top, which invigorates the lower, interior, shaded, weaker portions.

1.4.3 Root pruning:-

Root pruning is mostly removal of a part of the root system by cutting or heading back the end of the roots, but there is usually some thinning out too. Like top pruning, root pruning affects both the total growth of the plant and its vegetative-reproductive balance. In general, pruning the roots reduces the area for the absorption of nitrogen and other essential elements, and water. Growing-point cells are reduced as are cell division and enlargement, and there is reduced utilization and accumulation of carbohydrates. Reduction of these processes favors reproductive over vegetative processes.

Major effects of root pruning are:-**(i) reduced absorption, (ii) more branched main roots and feeder roots, (iii) reduced top growth.** Since reduced absorption promotes dwarfness, root pruning was a standard practice for dwarfing plants. Since carbohydrates are stored in the primary and secondary roots of woody plants, root pruning removes some of the carbohydrates, which would otherwise be used for growth. Root pruning is especially helpful in transplanting evergreen plants like citrus, banana or palm trees. It is also employed in the field with established plants to check over-vigorous growth, which may hasten bearing and increase productivity, but also may produce undesirable results.

In this respect, the effect of root pruning on apple trees was studied by **Khan et. al. (1998)** when they pruned the root to depth of 30 cm with sharp spade along both sides of the row and between trees, 20 or 30 cm from the trunk. They found that root pruning decreased tree height, shoot length, branch number, and shoot diameter compared with control, also root pruning increased the number of flowering spurs in the subsequent season altered. Root pruning reduced total yield and average fruit size.

1.5   Special pruning practices

1.5.1 Dehorning:-

Severe cutting back of trees and shrubs to keep them from becoming too tall is called dehorning. A somewhat less drastic heading back is employed with too-tall fruit trees to make spraying and harvesting easier.

1.5.2 Pinching:-

Removal of a plant part that is soft enough to be readily broken with the fingers is called pinching. One may remove a shoot or spur at its base (thinning) or remove its terminal portion (heading back) by pinching. If pinching is done early in the season it stops the growth of the pinched shoot and causes it to branch; if done late in the season it stops lengthening without stimulating branching.

1.5.3 Disbudding:-

Removal of buds is easier after they have started to break. Disbudding is useful in preventing the development of laterals on the trunks of recently set trees and sometimes in limiting the number of laterals on their main branches. However, the labor involved may often be greater than the benefits derived.

1.5.4 Stripping, notching, binding, ringing, and girdling:-

They are limited to treatment of the bark of the trunk or main branches and serve to restrict the downward flow of food from leaves to roots, resulting in the accumulation of food reserves above the point of restriction. This encourages fruit-bud differentiation, flowering, and fruiting. Tying knots in the flexible stems of young seedlings has an effect similar to that of girdling. Bending and tying down long vigorous shoots on a young tree in its formative years also has a similar effect. These operations should be regarded as special practices rather than ones to be regularly employed.

1.5.5. Fruit thinning:-

Removal of blossoms, tiny fruits, buds and spurs are ways of accomplishing fruit thinning. Pruning some fruiting branches prior to bloom will decrease the number of fruit. More shoot growth will occur, more carbohydrates will be manufactured, and the remaining fruit will be larger. Also, if adequate nutrients are available, increased food production will result in an abundance of fruit buds for the following year's production. Thus, fruit thinning in years when an excess of fruit buds are produced will help to overcome the biennial bearing habit of fruit and may force a biennial bearer to become annual.

With all fruit thinning, whether mechanical or chemical, it is important to remove the excess fruit while it is very small, less than 1.5 cm in diameter. If it is thinned after it gets larger than 2.5 cm in diameter, the pattern of fruit development has already been well established, and there may be little beneficial effect on the current crop (**Malik, 2000).** In this respect, several studies were done on the effect of thinning on yield of different fruit crops. However, **Francesconi, *et.al* (1996), Eman (1997), Naor *et.al.* (1997) and Koike& Ono (1998),** using a hand thinning on apple trees and found that if crop load sufficiently heavy, flower initiation well be inhibited, the optimum crop load of tree was related to leave/ fruit ratio. They add that the volumetric relative growth rate increased with decreasing crop load, also increasing crop load showed fruit weight reduction earlier and more severely than lightly cropped trees.

1.6 Chemical pruning

The use of chemicals in the regulation of plant growth is becoming more widespread because manual pruning is time-consuming and also very expensive. Chemicals are used for the following purposes.

- To increase the number of new shoots per plant.
- To control the shape of the plant.
- To control the number of flowers and fruits.
- To control the time of flowering.

Root pruning is also done with chemicals. Often a dominant tap root develops if the root is not pruned to induce strong, fibrous growth. Since hand pruning takes much time, chemicals are used to treat layers in the seedbeds. When the roots reach these layers, they are chemically pinched.

Plant height can also be controlled chemically. Since chemical used for this purpose have the desirable effect of reducing internode length without reducing the number of leaves.

In this respect, **Kabbel *et. al.* (1999), David & John (2000) and (2001)** used chemical thinning on apple trees and reported that the time of fruit thinning is very important, that carbaryl reduced fruit set when applied at 10- 12mm fruit diameter, but failed to thin at petal fall. They also reported that ethrel, NAA, hydrogen cynamide and hand thinning treatments were effective to thinning fruits of apple and improved fruit quality.

Fruit thinning are very important for peach, since trees loaded massive amounts of fruits. However, **Ezz, Thanaa & Amal (2000)** used GA_3 , urea at pre-bloom stage and ethrel at post-bloom stage of Early Grand peach trees, they found a significant reduce in yield but increase in fruit weight when used the previous treatment comparing with control, also the treatments increased TSS and vitamin C. Ethrel decreased chlorophyll but increased carotene and anthocyanin.

1.7   Amount of pruning related to bearing habits

The amount of pruning varies with plants of different ages. In general, after planting the aim should be to build up a proper framework. After the framework is built up, the next objective is to produce fruiting wood. Thus at first when plants are young, severe pruning is done to build up a strong framework. As the plant approaches bearing age, pruning is comparatively less severe to permit production of leaves and fruiting wood. In some cases even no pruning is desirable. With fruit crops like apples and pears, heavy pruning at this time will greatly delay fruiting. The severity of pruning increases as the tree increases its bearing area. In older trees, there is often too much bearing wood for the roots and leaf area to support. At this time old unproductive wood is pruned away to make the plant bear regularly.

The severity of pruning also varies for different fruit trees depending on their habits of fruiting. For example, grapes bear only on current season wood, so for them pruning is really an annual rejuvenating process. It removes all the vine except enough one-year-old wood to produce a sufficient number of fruiting shoots and leaves the minimum amount of old wood that will make a satisfactory framework for the vine. Thus each year about 90% of the vine is pruned out. In this respect, **Sourial *et.al.* (1999)** worked on 9 grapevine varieties and studied the effect of pruning treatments within a uniform bud load =35-36 bud/vine. The tested treatments were :(18 spurs each comprised 2 bus), (12 spurs x 3 buds), (9 spurs x 4 buds), (7 spurs x 5 bus) and (6 spurs x 6 buds). They found that the highest yield/vine of the tested varieties was obtained from the following combinations:- Cardinal (18 x 2), Black Rose (18 x 2), Ruby seedless (18 x 2), Exotic (18 x 2), Emerald seedless (9x4), Beauty seedless (18 x 2),Queen (6 x 6), Perlette (7 x 5) and Delight (all pruning treatments). On the other hand, **Omar & Abdel-Kawi (2000)** studied the effect of different bud load, namely 48,

60, 72, 90 and 120 buds per vine, they found that the optimal bud load/vine was 72 to 90 buds to produce high yield with good quality and maintaining the vigour of the vine.

With peaches, fruit is borne laterally on long vigorous one-year-old wood. Thus one-year-old wood must be developed for fruit production, and healthy older wood maintained for the production of satisfactory new fruiting wood. For regular production it is necessary to provide for an adequate annual supply of suitable shoots. Moderately vigorous trees are necessary to obtain such relatively thin fruit-bearing twigs. Pruning also serves as a way of fruit thinning. Often $^1/_3$ or more of the peach tree is pruned out. In this respect, **Bradley and Dagmar (1998)** studied the effect of timing and concentration of dilute foliar sprays of gibberellic acid on flower bud thinning of peach trees, they found that there was a strong trend for June sprays to minimize the total flower bud density, while treatments applied between early July and late October did not affect. Increasing the concentration from 25 or 50 mg·L^{-1} to 200 mg·L^{-1} tended to decrease total flower bud density of short shoots (<10 cm), but only about two-third as effectively as on long shoots.

Apple trees bear their fruit chiefly on old spurs distributed throughout the tree. The object of pruning is to maintain an adequate supply of sufficiently vigorous spurs. This is accomplished by a light thinning throughout the tree. A heavy pruning like that appropriate for peach trees would force these spurs into long vegetative growth and fruiting wood would be decreased. With apple trees, about $^1/_3$ or less of the fruiting wood is removed. Fruit-bearing habits and the amount of pruning desirable for various fruit plants are given in Table (1) as recorded by **Malik (2000)**:-

Table (1): **Fruit-bearing habits and amount of pruning in different fruit plants**

Fruit	*Bearing habits*	*Amount of pruning*
Apple, Pear	Fruit buds are mixed and on opening produce flowering, fruiting shoots with terminal inflorescence. They are borne on spurs of varying ages and occasionally on long shoots.	About $^{1}/_{3}$ or less fruiting wood is removed after harvesting the fruit.
Peach	Almost all fruit buds have flower parts only and are borne laterally on one-year-old shoots.	About $^{1}/_{3}$ or more of peach tree is pruned every year.
Almond, Plum, Apricot	Fruit buds have flower parts only and are present laterally both on spurs and shoots.	Less than $^{1}/_{3}$ of the fruiting wood is removed.
Grapes	Flowers are borne laterally on new growth of the current season which arises from lateral buds on the canes.	Severe pruning is done every year.
Mango	Fruit buds are borne terminally on long shoots.	Some pruning is done to control alternate bearing and remove old, undesirable.
Citrus	Flower buds are present on new growth appearing in spring, laterally or terminally.	Regular yearly pruning is essential only for lemons. For other citrus fruits only undesirable and diseased dry branches are removed.
Banana	Inflorescence is lateral and once in the life of the plant.	Whole plant is removed after harvesting the fruit.
Pome-granate	Flowers are borne terminally on spurs located on older peripheral wood. Spurs usually stop bearing after 2-3 years and new spurs are formed.	Light thinning out is practiced and no heading back is done.
Guava	Fruit buds are mixed and are terminal on one-year-old shoots. The buds on opening produce leafy shoots on which flowers are borne in the axils of the outermost leaves.	Only dry diseased undesirable parts are removed.
Olive	The bearing habit is similar to guava but the inflorescence is in the axils of the lower instead of the outermost leaves. The flowering shoots appear from lateral or terminal buds.	Only dry diseased undesirable parts are removed.
Fig	Bears lateral fruit buds.	After 3-4 years severe pruning is done.

2.   Population adjustment

One of the main targets that the horticulturists want to do is how to get the maximum yield and fruiting area among the planting unit in the shortest time?

In this respect, **Robinson *et.al.* (1997),** reported that the key objective in planning a new apple orchard should be to maximize yield in the early years and still effectively produce larger yield of high quality fruits in the second 10 years. They add that the best way to obtain high early yields is to plant a high density.

However, the effect of population adjustment or the other factors related to it (plant distribution and planting density) on fruit trees productivity was studied by many researchers with different fruit crops. On the other hand, there are many systems to distribute fruit trees through the plantation i.e. triangle, rectangle, square systems...etc. Moreover, the mechanical practices of the orchard are very affecting by the spaces between trees (between rows and within the row).

On the other side, presence of pollinator trees through the orchard is very important and limiting factor to obtain economical yield in many cultivars of fruit trees and to increase fruit set, consequently final yield production. Different methods of pollinator distribution can be used as follows:-

a) Complete along rows, as one row of the pollinator per two or four rows of the main cultivar according to the value and efficiency of the pollinator.

b) The second tree in the second row.

c) The third tree in the third row.

d) Across rows through the orchard.

However, there are many studies on different fruit trees cultivars were done and showed the importance of population adjustment either as planting system (plant distribution) or planting density (planting distance). We will take banana and apple as a sample for both evergreen and deciduous fruit trees to appear the effect of population adjustment on their productivity.

As for banana plants, **Simmonds (1982)** reported that spacing is a matter of extreme complexity. Practice in various parts of the world ranges very widely within the limits, 150- 2000 plants per acre (roughly 300-22 square feet per plant). Within these limits it is probably true to say that medium spacing of about 60-120 square feet per plant predominate. At least nine factors affecting the choice of spacing may be distinguished as follows:-

First:- the clone to be grown is important, when large plants in general being allowed more space than small ones.

Second:- spacing is affecting by soil fertility and by the fertilizing regime to be adopted. In general, the more fertile the soil and the heavier the fertilizing, the closer may the bananas be planted.

Third:- the pruning regime, inasmuch as it determines the effective population of the field, is very important. By allowing each stool to carry two bearing plants, the effective population is doubled and the planting distance therefore virtually halved. **Fourth:-** economic factors have their effect on the choice of spacing. If fruit is to be bought by weight the planter will normally be concerned to achieve high gross yields of fruit and will therefore choose dense planting.

Fifth:- various aspects of management affect the layout to be adopted although not the density of planting; thus, where implements or spraying machinery must be moved in the field and where irrigation water is to be applied, the plants will generally be laid out in rows rather than on the square ("hedgerow planting" in Jamaica). For example, if a mean density of about 400 plants per acre is appropriate, it may be better to space the plants 7 ft. apart in rows 14 ft. apart than on the square, 10 ft. x 10 ft.

Sixth:- densely planted bananas, by shade and nutrient competition, suppress weed growth to an extent that may materially reduce the cost of weed control.

Seventh:- closely planted bananas are generally believed to be more resistant to wind damage than widely planted ones. **Eighth:-** topography affects layout; on slopes, contour rows are preferable to square planting.

Ninth:- the history of the land may also affect layout; existing drainage systems and the presence of tree stumps may impose various restrictions and irregularities.

On the other hand, **Nakasone and Paull (1998)** indicated that planting densities of 1000-3000 plants ha^{-1} are used, with rectangle, single- or double- row cropping. Double rows (2m) combine higher density with a 3.5m alley for access. The actual density depends upon cultivar and climate. Higher densities are used in hot, dry localities to generate the necessary shade and microclimate for maximum yields. The vigour of a plantation is related to the canopy characteristics, leaf- area index and yield. If only one growth cycle is to be used, higher density may be planted (3000 plants ha^{-1}) , for three or more growth cycles, a lower density is recommended (2000 plant ha^{-1}) for highest gross margin.

However, the effect of planting system was studied by **Adjei _et.al._(1996)** who reported that plantain cultivar Borodewuio planted at 30 or 60 cm, in a square, triangular or rectangular arrangement at a density of approximately 1000 or 2500 plants/ha. Of the 3 factors investigated, population density had the most influence on plantain performance. Plants at the lower density had larger girths at flowering than those at the higher density, more leaves/plant and fewer broken pseudostems. They add that yields per plant were higher at the lower density (8.7 kg) than at the higher density (7.1 kg), but yields per ha were lower (9.4 vs. 17.0 t, respectively). Planting arrangement and depth had no significant effects on these parameters.

Apshara & Sathiamoorthy (1999) studied the effect of 3 spacings (2x2, 2x2.5 and 2x3m) and 3 planting densities (1,2 and 3 suckers/hill) on growth and yield of Nendran banana, they found that increasing planting density significantly increased pseudostem height and reduced pseudostem girth in all spacings, they also found that increasing plant density in all spacings decreased bunch weight, but increased total yield production. The highest yield was obtained with 2x2m space + 3 suckers/hill.

Moreover, **Nalina _et.al._(2000)** studied the effect of high density planting (HDP) on banana cv. Robusta (AAA). They reported that the treatments included conventional single sucker/pit, 3 and 4 suckers/pit at a spacing of 1.8m x 3.6m and 3.6m x 3.6m, they found that all (HDP) treatments resulted in increased pseudostem height, leaf number, leaf area, but reduced pseudostem girth.

Regarding the effect of population adjustment on apple trees, **Kron _et. al._ (2001)** worked on apple trees and estimated the magnitude of pollen dispersal across rows and along rows in two high density. They found that the maximum pollen dispersal distance was greater across rows (62.4 m) than along rows (13.7 m). However, the average dispersal distance across rows, expressed in meters or trees, (17.4 m and 3.6 trees) did not differ from that along rows (5.8 m and 2.7 trees).

On the other hand, **Zimmerman & Steffens (1995)** reported that the tree size and flowers amount of both Gala and Triple Red Delicious apple trees grown at three planting densities were inversely related to planting density. Also yield per tree decreased as tree density increased. They add that increasing tree density did not increase yield/ha.

Robinson *et. al.*(1997) on apple trees reported that the high yield per acre at maturity requires high total light interception. This can be accomplished by ensuring that row spacing is not too wide and that tree height is at least 75% of row spacing. They add that high fruit quality requires good light distribution throughout the tree canopy.

Moreover, **Wagenmakers & Callesen (1995)** worked and studied the effect of tree density (2000, 2667 or 4000 trees/ha) and the ratio of between to within-row distance (1:1, 2:1, and 3:1) on light interception, fruit production, colour and individual fruit weight of apple trees. They found that plantings with 1:1 and 2:1 between to within row distances intercepted more light and had a more uniform light distribution than 3:1 designs. This led to higher fruit production and better fruit colour. Fruit weight was not influenced by tree density, rectangularity or tree height.

On the other side, **Hrotko** *et.al.*(1999): reported that apple trees were planted in rows 5m apart and at the recommended distance for its rootstock or at spacings 15-20% less and greater than normal within rows. In the first 4 years, in-row spacing had no effect on growth or yield, as the trees had not filled their allotted space. However, yield/ha was greatest at the closest spacing.

<u>REFERENCES</u>

Adjei,N.S.; Dzietror, A.; Ahiekpor,E.K.S.; Afreh,N.K.; Ofori,I. and Hemeng,O.B. (1996): Effect of planting depth, density, and arrangement on growth, lodging, and yield of plantain (Musa AAB group). MusAfrica., No. 9, 9-11.(c.f. CAB abst. a.n. 960312322).

Apshara, S.E. and Sathiamoorthy, S. (1999): Effect of planting density and spacing on growth and yield of banana cv. Nendran (AAB).
South Indian Horticulture. 47: 1-6, 1-3. (c.f. CAB abst. a.n. 20013016691).

Radley, H.T. and Dagmar, G.T. (1998): Flower bud thinning and winter survival of "Redhaven"and "Cresthaven"Peach in Response to GA$_3$ sprays.
J. Amer. Hort. Sci. 123 (4): 500-508.

Caruso, T.; Inglese,P.; Sottile, F. and Marra, F.P. (1999): Effect of planting system on Productivity, dry-matter partitioning and carbohydrate content in above-ground components of "Flordaprince" peach trees.
J. Amer. Soc. Hort. Sci., 124(1): 39-45.

David, C.F. and John, C.S. (2000): Chemical thinning "Fuji" apple in the Midwest.
J. Amer. Pomo. Soc., 54 (2) : 61- 67.

 David, C.F. and John, C.S. (2001): Chemical thinning "Gala" apple in the Midwest
J. Amer. Pomo. Soc., 55 (2) : 109-113.

Eman, S.A. (1997): Effect of ways of hand thinning on ANNA apple fruit quality and vegetative growth under Egyptian desert conditions.
Egypt. J. Agric. Res., 76 (2) 693.

Francesconi, A.H.D.; Lakso, A.N.; Nyrop, J.P.; Barnard, J. and Denning, S.S. (1996): Carbon balance as a physiological basis for the interactions of European red mite and crop load on "Starkrimson Delicious" apple trees.
J. Amer. Soc. Hort. Sci., 121: 5, 959- 966.

Gaser, Aisha, S.A. and Osman, M.H. (1999): Effect of training system on yield and fruit quality of Flame seedless and Ruby seedless grapevine cultivars.
J. Agric. Sci. Mansoura Univ., 24 (12) 7559-7565.

Hortko, k.; Hanusz, B.; Magyar, L. and Sadowski, A. (1999): Effect of rootstocks and in-row spacing on growth and yield of "Idared" apple trees.
Proceedings of the international seminar, Warsaw-Ursynow, Poland, 18-21 August, 1999, 37-38.(c.f. CAB abst. a.n. 20000305373).

Kabbel, H.; Bahlod, S.E. and Bothina Abd El-Ghaffer (1999): Comparative studies on "ANNA" apple thinning.
Minufiya J. Agric. Res., 24 (4) 1321-1332.

Khan,Z.U.; McNeil, D.L. and Samad, A. (1998): Root pruning of apple trees grown at ultra-high density affects carbohydrate reserves distribution in vegetative and reproductive growth.

New-Zealand J. Crop and Hort. Sci., 26 (4) 291- 297. (c.f CAB abst. a.n. 990301435).

Koike, H. and Ono, T. (1998): Optimum crop load for Fuji apples in Japan.

Compact, fruit, tree. 31: 13-16. (c.f. CAB abst. a.n. 980303902).

Kron, P.; Husband, B.C. and Kevan, P.G. (2001): Across- and along- row pollen dispersal in high-density apple orchards: insights from allozyme markers.

Journal of Hot. Sci. and Biotech. 76: 3, 286-294.

Malik,M.N (2000): Horticulture, 231-244.

Biotech Books, Delhi.

Michele, R.W. (1997): The vertical axis system : A training method for growing apple trees.

Department   of   Horticulture,   Univ.   of   Missouri-   Columbia. http://muextension.missouri.edu/xplor/agguides/hort.

Nakasone, H.Y. and Paull, R.E. (1998): Tropical fruits., 112-120.

CAB International.

Nalina, L.; Kumar, N. and Sathiamoorthy, S. (2000): Studies on high density planting in banana cv. Robusta (AAA). 1. Influence on vegetative characters.

Indian Journal of Horticulture. 57: 3, 190- 195.

Naor, A.; Klein, I.; Doron, I.; Gal, Y. Ben, D. Z. and Bravdo, B. (1997): Irrigation and crop load interactions in relation to apple yield and fruit size distribution.

J. Am. Soc. Hort. Sci., 122: 3, 411- 414.

Omar, A.H. and Abdel- Kawi (2000): Optimal bud load for Thompson seedless grapevines.

J. Agric. Sci. Mansoura Univ., 25 (9) 5769-5777.

Retamales, J.B.; Lepe, V.P.; Herregods,M.; Boxus,P. Baets,W. and Jager,A. (2000): Control strategies for different bitter pit incidences in Braeburn apples.

Acta Horticulturae. No. 517, 227-233.

Robinson, T.L; Hoying, S.A; Smith, W.H. and Bramlage,   W.J. (1997): Training strategies for high density orchards of the

future.

New-England-Fruit-Meetings., 103: 27-36. (c.f. CAB abst. a.n. 980311757).

Shaffer, W.H. (2000): Fire blight.

Department   of   Plant   Pathology,   Univ.   of   Missouri-   Columbia. http://muextension.missouri.edu/xplor/agguides/hort.

Simmonds, N.W. (1982): Bananas, 158- 203.

Longman

Sourial, G.F.; Al-Ashkar,R.A.; Marwad, I.A.; Hassan, A.S. and Safaa, A. N. (1999): Productivity of some newly introduced grape varieties in relation to bud fertility and some pruning treatments.

Zagazig J. Agric. Res., 26 (3A) 657-672.

Ezz, Thanaa, M. and Amal, M.E. (2000): Response of Early Grand peach to chemical thinning; ethrel, gibberellic acid and urea. 1. yield and fruit quality.

J. Agric. Sci. Mansoura Univ., 25 (8) 5267- 5278.

Wagenmakers, P.S. and Callesen, O. (1995): Light distribution in apple orchard systems in relation to production and fruit quality.

J. Hort. Sci., 70: 6, 935-948.

Zimmerman, R.H. and Steffens, G.L. (1995): Cultivar, planting density, and plant growth regulator effects on growth and fruiting of tissue cultured apple trees.

J. Amer. Soc. Hort. Sci., 120 : 2, 183-193.